Aliyou NDAM

PAUL BIYA: Misunderstood architect of the agricultural revolution

Aliyou NDAM

PAUL BIYA: Misunderstood architect of the agricultural revolution

ScienciaScripts

Imprint

Cover image: www.ingimage.com

This book is a translation from the original published under ISBN 978-620-6-71291-6.

Publisher:
Sciencia Scripts
is a trademark of
Dodo Books Indian Ocean Ltd. and OmniScriptum S.R.L publishing group

120 High Road, East Finchley, London, N2 9ED, United Kingdom
Str. Armeneasca 28/1, office 1, Chisinau MD-2012, Republic of Moldova, Europe
Printed at: see last page
ISBN: 978-620-7-62954-1

MAP OF THE BOOK

PREAMBLE

The book "Paul Biya, misunderstood architect of the agricultural revolution" is based on several key elements. Firstly, it is important to note that agriculture plays a crucial role in Cameroon's economy, as a source of income for many citizens and as a key sector for the country's food security. However, despite its importance, the agricultural sector in Cameroon has often been criticised for its lack of modernisation, efficiency and government support.In this context, President Paul Biya has launched several initiatives and programmes to promote agricultural development in the country. These efforts include investing in agricultural infrastructure, promoting sustainable agricultural practices, putting in place policies to support farmers and seeking international partnerships to strengthen the sector.However, despite these actions, Paul Biya's agricultural policy has provoked controversy and criticism. Some observers believe that the reforms undertaken are not ambitious or effective enough, while others question the transparency and real impact of these initiatives on the ground.It is in this complex context that the book "Paul Biya, architecte de la révolution agricole incompris" seeks to analyse and shed light on the role of the Cameroonian president in transforming the agricultural sector. By critically examining his actions, motivations and results, the author invites readers to gain a better understanding of the issues surrounding agriculture in Cameroon and to question the perceptions surrounding Paul Biya's agricultural policy.

CHAPTER 1

AGRICULTURAL CONTEXT IN CAMEROON

Cameroon's agricultural sector plays a crucial role in the country's economy and society. Agriculture employs around 60% of the working population and makes a significant contribution to national GDP. However, despite its importance, Cameroon's agricultural sector faces a number of challenges that hamper its development.

The current agricultural situation in Cameroon is characterised by low productivity, traditional farming practices, excessive dependence on climatic conditions, inadequate infrastructure and limited access to agricultural inputs and markets. These factors limit the capacity of the agricultural sector to meet the food needs of the population and generate sustainable income for farmers.

The challenges facing Cameroon's agriculture also include deforestation, soil degradation, poor use of modern agricultural technologies, lack of training for farmers and vulnerability to external shocks such as climate change and health crises.

Despite these challenges, agriculture remains an essential pillar of Cameroon's economy, providing jobs, income and essential foodstuffs for the population. Promoting agricultural development is therefore crucial to boosting food security, reducing rural poverty and stimulating the country's economic growth.

A- History of the agricultural revolution in Cameroon under the Paul BIYA regime

The history of Cameroon's agricultural revolution under Paul Biya is marked by a series of initiatives aimed at modernising and boosting the country's agricultural sector.

1. Historical background

- Paul Biya became President of Cameroon in 1982, succeeding Ahmadou Ahidjo. As soon as he came to power, he introduced reforms aimed at modernising the Cameroonian economy, including the agricultural sector.

2. The main thrusts of the agricultural revolution

- Crop diversification: Under the Paul Biya regime, efforts have been made to encourage the diversification of agricultural crops, with an emphasis on food crops and export crops.

- Agricultural mechanisation: The government has also invested in agricultural mechanisation to improve the efficiency and productivity of farms.

- Promoting agro-industry: In order to enhance the value of local agricultural products, measures have been taken to promote the development of agro-industry and agri-food chains.

3. Challenges encountered

-Despite the efforts made, Cameroon's agricultural revolution under Paul Biya has faced a number of challenges, such as low farm productivity, difficulties in accessing agricultural inputs and markets, and the effects of climate change.

- Persistent inequalities in the distribution of land and agricultural resources have also limited the progress of the agricultural revolution.

4. Future prospects

- To meet these challenges, it is essential to strengthen policies and programmes aimed at supporting small-scale farmers, improving access to agricultural inputs and markets, and promoting sustainable and resilient agriculture.
- Collaboration with local stakeholders, including civil society organisations and international partners, is essential to ensure the success of the agricultural revolution in Cameroon.

Cameroon's agricultural revolution under Paul Biya has been marked by efforts to modernise and boost the country's agricultural sector.

To ensure the sustainability and prosperity of Cameroon's agriculture, it is crucial to continue investing in policies and programmes tailored to farmers' needs and to promote inclusive and sustainable agriculture.

B - Structuring the world of agriculture

The structure of the agricultural world in Cameroon is characterised by a diversity of agricultural production systems, regional disparities and a high level of dependence on small-scale farmers.

-Cameroon is a Central African country with great climatic and geographical diversity, which translates into a variety of agricultural production systems.

- Agriculture plays a central role in the country's economy, employing the majority of the working population and making a significant contribution to GDP.

1. The main agricultural production systems

- Subsistence farming: practised by many smallholders, this aims to ensure food security for local populations.

- Commercial agriculture: developed in more favourable regions, this aims to produce export crops such as cocoa, coffee, cotton and oil palm.

- Livestock farming: practised in the Sahelian regions of the north, this is a major source of income for many livestock farmers.

2. Regional disparities

- The regions from north of the country are characterized by more arid climatic conditions, limiting the scope for agricultural production.

- The more humid southern regions are ideal for growing export crops such as cocoa and coffee.

3. The dependence of smallholders

- The majority of farmers in Cameroon are small-scale producers, often facing difficulties in accessing agricultural inputs, markets and finance.

- The government's agricultural policies aim to support these small farmers through training programmes, access to credit and the development of agricultural infrastructure.

The structure of the agricultural sector in Cameroon is marked by a diversity of production systems, regional disparities and a high level of dependence on smallholders. To ensure the sustainability and prosperity of the agricultural sector, it is essential to put in place policies and programmes tailored to the specific needs of the various players in the agricultural world.

C - The strengths of Cameroon's agricultural sector

Cameroon's agricultural sector has many assets that make it a pillar of the country's economy. Here are a few key points to highlight:

1. **Diversity and wealth of agricultural resources**: Cameroon boasts a wide variety of climates and soils, enabling the production of a wide variety of agricultural crops, ranging from staple crops such as wheat and maize, to maize and rice. maize, cassava and rice, to export crops such as cocoa, coffee and cotton.

2. **Agricultural production potential:** Cameroon has vast tracts of arable land, offering significant potential for increasing agricultural production. In addition, the country has an abundant and skilled workforce in the agricultural sector.

3. **Contribution to the national economy:** The agricultural sector is a major contributor to Cameroon's economy, accounting for around 20% of the country's GDP. It also provides employment for a large proportion of the working population, particularly in rural areas.

4. **Export potential:** Cameroon is a major exporter of agricultural products, particularly cocoa, coffee and cotton. These products make a significant contribution to the country's export earnings and strengthen

its position on international markets.

5. Favourable investment and policies: The Cameroonian government has put in place policies and programmes to promote the development of the agricultural sector, such as tax incentives for agricultural investment, subsidies for farmers and measures to improve rural infrastructure.

Cameroon's agricultural sector has many assets that make it a key sector for the country's economic development. With its abundant natural resources, high production potential, contribution to the national economy and favourable policies, Cameroon has everything it takes to continue strengthening its agricultural sector and making it an engine for sustainable growth.

CHAPTER 2

PRESIDENT PAUL BIYA'S AGRICULTURAL POLICY

President Paul Biya's agricultural policy in Cameroon aims to promote the sustainable development of the agricultural sector by putting in place measures to improve agricultural productivity, strengthen food security, reduce rural poverty and stimulate economic growth.

A - Guiding principles of Paul Biya's agricultural policy

The guiding principles of Paul Biya's agricultural policy in Cameroon are as follows:

1. **Promoting food security**: Agricultural policy aims to guarantee access to adequate and nutritious food for all Cameroonians by promoting local production and reducing dependence on food imports.

2. **Sustainable development**: Paul Biya's agricultural policy emphasises the promotion of environmentally-friendly agriculture, which preserves natural resources and helps to mitigate the effects of climate change.

3. **Strengthening agricultural productivity**: Agricultural policy aims to improve farm productivity by introducing modern technologies, providing access to quality inputs and building farmers' capacity.

4. **Promoting commercial agriculture**: The government is encouraging the development of commercial agriculture by facilitating farmers' access to markets, promoting the processing of agricultural products and supporting initiatives to add value to local products.

5. Social inclusion: Agricultural policy aims to reduce rural poverty by promoting the social inclusion of farmers, particularly smallholders, women and young people, through training programmes, access to credit and technical support.

6. Public-private partnerships: Agricultural policy encourages collaboration between the public sector, the private sector and civil society to promote the development of the agricultural sector, by encouraging private investment and strengthening partnerships for rural development.These guiding principles guide the actions of the Cameroonian government in the agricultural sector in order to promote sustainable, inclusive and resilient development of the agricultural sector in Cameroon.

B - Political measures and guidelines and the very high instructions of President Paul Biya to promote agricultural development

President Paul Biya's political guidelines and very high instructions for promoting agricultural development in Cameroon are as follows:

1. Investment in agriculture: President Paul Biya has issued instructions to increase investment in the agricultural sector in order to modernise farms, improve rural infrastructures and increase agricultural productivity.

2. Promoting research and innovation: The President encouraged the strengthening of agricultural research and innovation to introduce modern technologies, improve farming practices and develop crop varieties adapted to local conditions.

3. Strengthening agricultural sectors: The government has received instructions to support the development of agricultural sectors by encouraging the processing of agricultural products, improving access to markets and promoting the certification of local products.

4. Promoting agricultural entrepreneurship: The President encouraged the promotion of agricultural entrepreneurship by supporting young farmers and facilitating access to finance, training and technical support services.

5. Capacity-building for farmers: The government has received instructions to build farmers' capacity in farm management, sustainable production techniques and the marketing of agricultural produce.

6. Promoting international cooperation: President Paul Biya emphasised the importance of international cooperation for agricultural development by promoting partnerships with other countries, regional organisations and international institutions.

These political guidelines and very high instructions from President Paul Biya aim to promote sustainable, inclusive and resilient agricultural development in Cameroon, by contributing to food security, reducing rural poverty and creating jobs in the agricultural sector.

C - Long-term objectives of Paul BIYA's agricultural policy

The long-term objectives of President Paul Biya's agricultural policy in Cameroon are as follows:

1. Ensuring food security: One of the main objectives is to guarantee food security for the population by increasing agricultural production, improving access to a healthy and balanced diet, and reducing dependence on food imports.

2. Promoting rural development: Agricultural policy aims to stimulate the economic and social development of rural areas by creating employment opportunities, strengthening basic infrastructure, improving access to basic social services and reducing inequalities between urban and rural areas.

3. Encouraging agricultural diversification: This involves promoting the diversification of crops, production systems and agricultural sectors to strengthen farmers' resilience in the face of climatic hazards, crop disease and market fluctuations.

4. Improving agricultural competitiveness: The policy aims to increase the competitiveness of the agricultural sector by modernising farms, improving product quality, building the capacity of players in the sector and promoting access to national and international markets.

5. Promoting environmental sustainability: The aim is to promote sustainable farming practices that preserve natural resources, reduce the environmental impact of agriculture and help mitigate climate change.

6. Strengthening governance in the agricultural sector: This involves improving governance in the agricultural sector by strengthening public institutions, encouraging the participation of local stakeholders and guaranteeing transparency and fairness in the management of agricultural resources.

These long-term objectives aim to transform Cameroon's agricultural sector to make it more productive, sustainable, inclusive and competitive, thereby contributing to the country's economic and social development and improving the living conditions of rural populations.

D- Paul Biya's main agricultural policy initiatives include :

a) - Some programmes and projects at Ministry at charge of Agriculture

1 - CEREAL STORAGE PROJECT:

*** Objectives:**

-Intervene on markets in cereal-producing and consuming areas to build up security and buffer stocks for

- fighting famine and poverty

- stabilise prices from one campaign to the next.

*** Support granted :**

- Construction of storage warehouses ;

- Rehabilitation of rural roads ;

- Locust and bird control and stockpile treatment ;

- Promoting community village storage.

2- RURAL DEVELOPMENT PROJECT FOR THE MOUNT MBAPPIT REGION

*** Objectives:**

- sustainably improve agricultural production and household incomes in Noun ;

- improve household food security in Noun ;

- ensure sustainable management of natural resources in the Noun region.

*** Support granted :**

- Construction of socio-economic infrastructure ;

- Development of low-lying areas ;

- Animation, training for beneficiaries and support for agricultural extension ;

- Farm inputs and equipment.

3 - NATIONAL FOOD SECURITY PROGRAMME

*** Objectives:**

- Increase crop, pasture and fish production by introducing adapted improved varieties and supplying inputs;

-Securing production by controlling water, managing soil fertility, protecting the environment and conserving natural resources;

- Improve the cash income of producers, especially women and young people;

- Improve the cereal storage system, particularly in high-risk areas;

-Contribute to improving the nutritional status of the population;

* **Support granted :**

- Development of lowlands ;

- Financing micro-projects;

- Agricultural inputs.

4-PROGRAMME FOR THE RENOVATION AND DEVELOPMENT OF VOCATIONAL TRAINING IN THE AGRICULTURE AND LIVESTOCK SECTORS (AFOP)

* **Objectives:**

- Provide training for producers and post-primary training for young people planning to set up in agriculture;

- Support for the installation of young farmers ;

* **Support granted :**

- Training ;

- Financing young people setting up in agriculture

5 -SUPPORT PROGRAMME FOR PROJECT MANAGEMENT BY RURAL SECTOR ADMINISTRATIONS

* **Objectives:**

- Strengthen the monitoring and evaluation capacities of the two ministries (MINADER and MINEPIA);

- Support the renovation of the agricultural, pastoral and fisheries statistics system;

- Strengthen the capacities and logistics of particularly at the decentralised level;
- Develop dialogue with other players in the sector.

* **Support granted** (to rural sector administrations) :

- Support for regional service improvement programmes (PRAS);

- Equipping the departments of both Ministries;

- Support for the renovation and development of statistical systems.

6 - PROJECT OF RELAUNCH OF THE RIZICULTURE IN THE LOGONE VALLEY

* **Objective:**

- Increasing rice production to boost food security in this ecologically fragile region.

*** Support granted :**

- Improved seeds ;

- Fertilisers and plant protection products ;

- Shelling machines ;

- Storage warehouses

7- PADMIR: RURAL MICROFINANCE DEVELOPMENT SUPPORT PROJECT

*** Objectives:**

-Improving the institutional environment for microfinance

-Facilitating access for rural populations to financial services and products tailored to their needs.

*** Support granted :**

- Project financing ;

- Capacity building for microfinance institutions

8 -"PACA: THE AGRICULTURAL COMPETITIVENESS IMPROVEMENT PROGRAM"

Objective:

-improve the competitiveness of eligible farmer organisations working in targed crop sectors, as well as the income of farmers who are members of these associations.

*** What benefits from the projetct?**

- Rehabilitation of key rural infrastructure (rehabilitation of irrigate areas and rural roads);

- Economic partnerships (co financing of investment sub projects and assistance in building economic partnerships);

- Institutional support and capacity building.

9 - PROGRAMME OF DEVELOPMENT OF VILLAGE PALM GROVES

*** Objective:**

-Improve at sustainable thelevel of living of farmers by an increase in stable income from palm oil production.

*** Support granted :**

- Agricultural inputs ;

- Processing equipment

10 - PROGRAMME TO BOOST THE PLANTAIN SECTOR

*** Objectives:**

- Setting up a network of plantain nurseries through private professionals or town halls;

- Support for farmers interested in creating a plantation of 0.5 to 1 ha or more of pure crop.

* **Support granted :**

- Training in plant propagation techniques;

- Agricultural inputs ;

- Direct subsidies to nurseries and municipalities.

11 - NATIONAL SUPPORT PROGRAMME FOR THE MAIZE SECTOR

* **Objectives:**

-Providing support to build the capacity of producers;

-Support the structuring and organisation of the sector;

-Supporting capacity building for producers ;

-Facilitating the use of quality seeds.

* **Support granted :**

- Seeds and other agricultural inputs ;

- Agricultural equipment ;

- Tractors for medium and large farms;

12- PROJECT T O SUPPORT THE PROTECTION OF COCOA/COFFEE ORCHARDS

13- PROJECT TO COMBAT MAJOR FOOD CROP SCOURGES

14-PROGRAMME NATIONAL OF VULGARISATION AND AGRICULTURAL RESEARCH

15- DECENTRALISED RURAL CREDIT PROJECT

16-PROJECT SUPPORT AT ESTABLISHMENTS OF MICROFINANCE DEVELOPMENT INSTITUTIONS

17-PROJECT SUPPORT À INTEGRATION OF YOUNG PEOPLE IN AGRICULTURE

18- KENNEDY ROUND 2 PROJECT

19- PROJECT TO RELAUNCH THE POTATO SECTOR

20- FERTILISER SUB-SECTOR REFORM PROGRAMME

21- COCOA COFFEE SEED PROJECT

22- MUSHROOM INDUSTRY DEVELOPMENT PROJECT

23- RURAL DEVELOPMENT PROGRAMME FOR THE MOUNGO-NKAM AGRICULTURAL BASIN

24- COCOA AND COFFEE DEVELOPMENT FUND

25- NATIONAL PROGRAMME FOR THE MANAGEMENT OF OBSOLETE PESTICIDES IN CAMEROON

26- EMERGENCY PROGRAMME TO REDUCE PESTICIDE RESIDUES IN COCOA / COFFEE CAMEROON

27- PROJECT TO SUPPORT THE DEVELOPMENT OF AGRICULTURAL VALUE CHAINS

28- PROJECT TO SUPPORT THE PRODUCTION AND DISTRIBUTION OF IMPROVED COCOA/COFFEE PLANT MATERIAL

29 - PILOT PROGRAMME TO REVIVE THE COCOA/COFFEE/COTTON SECTORS

30 - PROJECT TO SUPPORT THE USE OF COCOA/COFFEE FERTILISERS

31 - SUPPORT PROJECT FOR FUNGAL CONTROL ON COCOA/COFFEE

32 - PROJECT TO DEVELOP RAINFED UPLAND RICE CULTIVATION IN BIMODAL RAINFALL FOREST ZONES

33 - DEVELOPMENT OF VILLAGE HEVEA PLANTATIONS

34 - DEVELOPMENT OF MEDIUM-SIZED AND LARGE FARMS

35 - DEVELOPMENT OF THE PULSES SECTOR

36 - PROJECTS TO SUPPORT THE NATIONAL SEED PROGRAMME

37 - PILOT IRRIGATED RICE FARM CAMEROON -COREE

38 - AREA PARTICIPATORY DEVELOPMENT PROJECT

39 - PROJECT TO SUPPORT THE AGRICULTURAL SECTOR

40 - AGRICULTURAL INVESTMENT AND MARKET DEVELOPMENT PROJECT

41 - IMPROVEMENT FROM THE COMPETITIVENESS OF AGRO-PASTORAL FAMILY FARMS

42 - AGRICULTURAL VALUE CHAIN DEVELOPMENT PROJECT

43 - COCOA DEVELOPMENT SUPPORT PROJECT

44 - COFFEE DEVELOPMENT SUPPORT PROJECT

45 - CASHEW DEVELOPMENT SUPPORT PROJECT

b) -Some programmes and projects of the Ministry in charge of investments

1 RURALDEVELOPMENT PROJECT INTEGRATEDCHARI LOGONE - PDRI-CL

2 -PROGRAMME OF DEVELOPMENT PROGRAM DEVELOPMENT PROGRAM - PDICA

3 -PROJECT IMPROVEMENT OF THE AGRICULTURAL COMPETITIVENESS IN CAMEROON - PACA

4 - AGROPOLES PROGRAMME - PA

5 - GRASSROOTS POVERTY REDUCTION SUB-PROGRAMME - SPRPB II

6 - PROJECT PLANNING AND FROM VALORISATION OF INVESTMENTS IN THE VALLEY FROM LA BÉNOUÉ - VIVA-BÉNOUÉ

7 - PROJECT PLANNING AND FROM VALORISATION OF INVESTMENTS IN THE LOGONE VALLEY - VIVA-LOGONE

Despite these initiatives, challenges persist in Cameroon's agricultural sector, such as low productivity, lack of adequate infrastructure, deforestation and soil degradation. It is therefore necessary to strengthen agricultural policies and programmes to meet these challenges and promote sustainable agricultural development in Cameroon.

E. Agriculture as an engine for economic growth :

Agriculture plays a crucial role as an engine of economic growth in President Paul Biya's agricultural policy in Cameroon. Here are a few key points that illustrate how agriculture is seen as an important lever for boosting the country's economic development:

1. **Contribution to the economy**: Agriculture is one of the main economic sectors in Cameroon, making a significant contribution to the country's gross domestic product (GDP). By focusing on agricultural development, the government aims to increase agricultural production, create jobs in rural areas and boost farmers' incomes.

2.**Job creation**: Agriculture offers employment opportunities for a large proportion of the working population in Cameroon, particularly in rural areas. By investing in the agricultural sector, the government is seeking to create more jobs in agricultural production, agri-food processing, the marketing of agricultural products, and so on.

3. **Poverty reduction**: By strengthening the agricultural sector, agricultural policy aims to reduce poverty by increasing farmers' incomes, improving their access to markets and promoting the economic development of rural areas.

4. **Food security**: By increasing agricultural production and promoting crop diversification, agricultural policy helps to strengthen the country's food security by ensuring an adequate supply of food products for the population.

5. Agricultural exports: Cameroon has significant potential for exporting agricultural products to regional and international markets. By supporting agricultural exports, the policy aims to boost export earnings, diversify the economy and strengthen the competitiveness of the agricultural sector.

Agriculture is seen as an engine of economic growth in Paul Biya's agricultural policy for Cameroon, because of its potential to create jobs, reduce poverty, strengthen food security and contribute to the country's economic development.

F. Challenges and opportunities for an emerging Cameroon by 2035 :

Cameroon's agricultural sector faces a number of challenges and opportunities as the country aims to become an emerging country by 2035. Here are some of the main challenges and opportunities facing the agricultural sector:

- Challenges :

1. **Climate change:** Climate change can have a negative impact on agricultural production by altering rainfall patterns, increasing temperatures and favouring the emergence of diseases and pests. It is essential to put in place adaptation measures to mitigate the effects of climate change on agriculture.

2. **Inadequate infrastructure**: Agricultural infrastructure such as roads, storage warehouses, markets and irrigation systems are often inadequate, limiting farmers' ability to access markets, store their produce and optimise production.

3. **Access to finance**: Many farmers in Cameroon find it difficult to access the finance they need to invest in their farms, buy agricultural inputs or modernise their equipment. Better access to finance is essential to boost agricultural productivity and promote the development of the sector.

- Opportunities :

1. **Crop diversification**: Cameroon has a diverse climate and fertile soils that offer opportunities for crop diversification. By encouraging farmers to grow a variety of crops, the country can reduce its dependence on a few main crops and boost food security.

2. **Investment in agricultural research**: Investment in agricultural research can help develop crop varieties that are resistant to diseases, pests and changing climatic conditions. By supporting agricultural research, Cameroon can improve agricultural productivity and environmental sustainability.

3. **Promotion of agro-industry**: The development of agro-industry can create jobs, add value to local agricultural products and stimulate economic growth. By encouraging agri-food processing and the creation of agricultural value chains, Cameroon can capitalise on its agricultural resources to generate additional income. Cameroon's agricultural sector faces major challenges, but also presents numerous opportunities to contribute to the country's emergence by 2035. By investing in sustainable solutions, building farmers' capacity and promoting innovation, Cameroon can realise its agricultural potential and achieve its economic development goals.

CHAPTER 3

THE REASONS FOR THE REPEATED FAILURE OF PRESIDENT PAUL BIYA'S AGRICULTURAL POLICY

A - Structural reasons

President Paul Biya's agricultural policy in Cameroon is marked by challenges and difficulties that hamper its effective implementation. Despite efforts to promote agricultural development, several factors contribute to the repeated failure of this policy.

1. Lack of adequate funding :

One of the main reasons for the failure of President Paul Biya's agricultural policy is the lack of adequate funding. The resources allocated to the agricultural sector have often been insufficient to meet farmers' needs in terms of investment, infrastructure and technical support. This lack of funding has limited the agricultural sector's ability to modernise and increase productivity.

2. Insufficient agricultural infrastructure :

The lack of adequate agricultural infrastructure, such as roads, storage warehouses and irrigation systems, has also contributed to the failure of President Paul Biya's agricultural policy. These infrastructures are essential to facilitate farmers' access to markets, to store their produce adequately and to improve their productivity. Inadequate infrastructure has hampered the development of the agricultural sector and limited opportunities for growth.

3. Lack of coordination and monitoring :

Another factor that has contributed to the failure of President Paul Biya's agricultural policy is the lack of coordination and monitoring of agricultural programmes and projects. Insufficient coordination between the various players in the In addition, inadequate monitoring of agricultural programmes and projects has made it difficult to assess their impact and effectiveness. In addition, inadequate monitoring of agricultural programmes and projects has made it difficult to assess their impact and effectiveness.

4. Impact of climate change :

Climate change has also had a negative impact on President Paul Biya's agricultural policy. Climatic variations, such as droughts and floods, have affected agricultural production, reduced yields and threatened food security. Lack of adaptation to climate change has made the agricultural sector more vulnerable and undermined efforts to promote sustainable agricultural development.

5. No agricultural land policy

The absence of a clear and effective agricultural land policy in Cameroon has played a major role in the failure of the agricultural revolution initiated by President Paul Biya. Here are some of the reasons why this has had a negative impact on the development of the agricultural sector:

- **Access to land**: Without a clear land policy, many farmers in Cameroon face difficulties in accessing arable land. Competition for arable land can be intense, particularly with the expansion of mining, forestry and industrial activities. This limits farmers' ability to develop their farms and invest in modern farming practices.

- **Security of tenure:** The absence of guarantees of security of tenure for farmers makes long-term investment in agriculture risky. Land disputes and litigation can discourage farmers from modernise their farms or adopt innovative farming techniques, for fear of losing their land.

- Rural **development**: An inadequate land policy can also hamper rural development by preventing the provision of essential agricultural infrastructure, such as roads, irrigation systems and warehouses. This limits farmers' ability to market their produce, access agricultural inputs and benefit from support services.

- **Sustainable management of natural resources:** An inadequate agricultural land policy can also contribute to the unsustainable use of natural resources, such as deforestation, soil erosion and water pollution. Without a solid regulatory framework to protect agricultural land and ecosystems, the sustainability of agriculture is compromised.

The absence of a coherent and effective agricultural land policy in Cameroon has created an uncertain and unstable environment for farmers, limiting their ability to contribute fully to the desired agricultural revolution. Land reform is therefore essential to support the sustainable development of the agricultural sector and improve the country's food security.

Several factors have contributed to the failure of President Paul Biya's agricultural policy in Cameroon. Lack of adequate funding, inadequate agricultural infrastructure, lack of coordination and monitoring, and the impact of climate change have all played a role in limiting the development potential of the agricultural sector. To overcome these challenges, it is essential to adopt more integrated approaches, increase investment in the agricultural sector and promote more efficient management of resources to ensure sustainable and inclusive agricultural development in Cameroon.

B - The troubled role of agricultural engineers

Agricultural engineers play a crucial role in the implementation of a country's agricultural policy, providing technical expertise and know-how to improve the productivity and sustainability of the agricultural sector. However, in the context of Cameroon and President Paul Biya's agricultural policy, the role of agricultural engineers has sometimes been controversial due to various factors.

1. Political influence and corruption :

In many cases, agricultural engineers have been influenced by political interests or corrupt practices that have compromised their objectivity and integrity in implementing agricultural policy. Some agricultural engineers may have favoured certain players in the agricultural sector in exchange for political or financial favours, to the detriment of the general interest and equitable development of the sector.

2. Lack of professionalism and competence:

Some agricultural engineers have also been criticised for their lack of professionalism and competence in carrying out their tasks. The lack of adequate training, rigorous monitoring of recommended agricultural practices and effective implementation of agricultural programmes and projects has hampered progress in the agricultural sector and contributed to the failure of President Paul Biya's agricultural policy.

3. Conflicts of interest and favouritism :

Agricultural engineers can sometimes be faced with conflicts of interest due to personal or professional links with certain players in the agricultural sector. Such favouritism can lead to biased decisions and actions that are not in line with the government's agricultural development objectives, thereby compromising the effectiveness of agricultural policy.

4. Lack of transparency and accountability:

The lack of transparency and accountability in the actions of agricultural engineers has also contributed to their murky role in the implementation of President Paul Biya's agricultural policy. The absence of effective monitoring and evaluation mechanisms has allowed some agricultural engineers to act with impunity, without being held accountable for their actions or decisions.

5- Lack of an association of agronomists

The inability of agronomists in Cameroon to come together around a professional body can have a negative impact on the implementation and success of agricultural policies in the country. Here are some of the ways in which this can contribute to the failure of agricultural policies:

- **Lack of coordination**: The absence of a unified professional body for agronomists can lead to a lack of coordination and collaboration between the various players in the agricultural sector. This can make it difficult to implement agricultural policies coherently and coordinate efforts to achieve set objectives.

-**Weak political influence**: Without a strong and unified collective voice, agronomists may find it difficult to influence policy-makers and advocate effectively for favourable agricultural policies. This can lead to a failure to take account of the needs and recommendations of agronomy professionals in policy-making.

- **Lack of professional standards:** The absence of a professional body can also lead to a lack of professional norms and standards in the agricultural sector. This can compromise the quality of agronomic services provided to farmers and limit the effectiveness of agricultural interventions implemented as part of agricultural policies.

-**Fragmentation of knowledge and skills:** The lack of collaboration between agronomists can lead to a fragmentation of knowledge and skills in the agricultural sector. This can limit the ability of agronomy professionals to share best practice, innovate and leverage complementary skills to solve complex agricultural challenges.

The inability of agronomists in Cameroon to come together under a professional body can weaken the agricultural sector as a whole, compromising the effective implementation of agricultural policies and the achievement of the country's agricultural development goals. It is therefore crucial for agronomy professionals to work together to strengthen their collective voice, promote high professional standards and make a significant contribution to improving the agricultural sector in Cameroon.

The role of agricultural engineers in the implementation of President Paul Biya's agricultural policy in Cameroon has been marked by problems such as political influence, corruption, lack of professionalism, conflicts of interest and lack of transparency. To ensure the success of agricultural policies and promote sustainable agricultural development, it is essential to establish a culture of professionalism, integrity and accountability among agricultural engineers and stakeholders in the agricultural sector.

C - The Ministry of Agriculture and Rural Development (MINADER): a graveyard of billions of euros

The Ministry of Agriculture and Rural Development is a key player in the promotion of agriculture and rural development in many countries. However, in Cameroon, this ministry has become a place where billions of public funds are spent without tangible results, due to the multiplicity of projects costing billions that are implemented. implementation. What's more, the outsourcing of virtually all the Ministry's activities to these projects, whose coordinators are often regarded as demigods, raises questions about the management of resources and the effectiveness of agricultural policies.

1. Multiplicity of projects costing billions :

The Ministry of Agriculture and Rural Development is the scene of a multitude of projects financed at great expense, often without effective coordination or a clear strategic vision. This dispersal of resources leads to significant financial wastage and inefficiency in achieving agricultural development objectives.

2. Outsourcing the Ministry's activities :

In many cases, the Ministry outsources the implementation of its activities through specific projects, entrusting responsibility to Coordinators who often have considerable discretionary power. These Coordinators can act as demigods, taking unilateral decisions without real oversight or accountability, which can lead to abuse and mismanagement of public funds.

3. Opaque management of resources :

The management of resources within the Ministry of Agriculture and Rural Development can sometimes lack transparency and accountability. Control and monitoring mechanisms may be insufficient to ensure effective and efficient use of public funds, which may encourage corruption and misappropriation of resources.

4. Impact on agricultural development :

Excessive outsourcing of the Ministry's activities and the dispersion of resources in a multitude of projects costing billions can have a negative impact on agricultural development. Coordination Inadequate planning, a lack of strategic vision and opaque management of resources can undermine efforts to improve agricultural productivity, reduce rural poverty and ensure food security. The Ministry of Agriculture and Rural Development can sometimes be perceived as a graveyard of billions because of the multiplicity of projects costing billions that are implemented there, the excessive outsourcing of activities to these projects and the discretionary power granted to the Coordinators. To ensure efficient use of public funds and promote sustainable agricultural development, it is essential to improve coordination, transparency and accountability within the Ministry.

D - Unsuitable approach

The agricultural extension approach, which involves disseminating agricultural knowledge and techniques to farmers in order to improve their productivity, has long been a pillar of agricultural development in Cameroon. However, with the advent of the Conseil Agricole, a new approach based more on individualised advice to farmers, the gradual abandonment of agricultural extension has had a negative impact on the implementation of the agricultural revolution conceived by President Paul Biya.

1. Importance of the agricultural extension approach :

For decades, the agricultural extension approach has made it possible to effectively disseminate agricultural innovations, improve farming practices and increase agricultural productivity in Cameroon. It has contributed to the training of farmers and the adoption of new technologies, playing a key role in the development of the agricultural sector.

2. Transition to Farm Advisory Services :

With the advent of Farm Advisory Services, a new, more individualised approach based on direct advice to farmers was promoted.This transition has led to a gradual disinvestment in traditional agricultural extension, in favour of a model more focused on personalised advice to farmers.

3. Negative consequences of abandoning agricultural extension :

The abandonment of the agricultural extension approach has had harmful consequences for the implementation of the agricultural revolution conceived by President Paul Biya. The dissemination of agricultural innovations has been hampered, farmers have not benefited from adequate support to adopt new practices and technologies, and agricultural productivity has not grown as expected.

4. Impact on agricultural development :

As a result, the agricultural sector in Cameroon has suffered from a lack of significant progress as part of the envisaged agricultural revolution. The low uptake of new techniques, the lack of training for farmers and the lack of support have limited productivity gains and hampered the sustainable development of the sector.

The gradual abandonment of the agricultural extension approach in favour of the Agricultural Advisory Service has had a detrimental impact on the implementation of Paul Biya's agricultural revolution in Cameroon. To relaunch the agricultural sector and achieve the development objectives set, it is essential to reassess the policies and strategies for supporting farmers and disseminating agricultural innovations.The split between agricultural research and extension has also had a significant negative impact on the implementation of Paul Biya's agricultural policy in Cameroon. Here are a few key points to illustrate this impact:

1. Disconnect between research and extension :

The disconnect between agricultural research, which aims to develop new agricultural technologies and practices, and agricultural extension, which aims to disseminate these innovations to farmers, has led to a disconnect between the scientific knowledge produced by researchers and its application in the field by farmers. This disconnect has limited the effectiveness of the dissemination of agricultural innovations and hampered progress in the agricultural sector.

2. Difficulty in adopting new technologies :

As a result, farmers have often found it difficult to adopt the new technologies and farming practices developed by researchers. The lack of communication and coordination between research and extension players has made it difficult to pass on knowledge and support farmers in adopting these innovations, which has limited productivity and efficiency gains in the agricultural sector.

3. Loss of development opportunities :

The disconnect between agricultural research and extension has also led to a loss of development opportunities for the agricultural sector in Cameroon. The scientific and technological advances developed by researchers have not been fully exploited in the field because of the lack of coordination and collaboration between the various players involved in the agricultural innovation process.

4. Impact on Paul Biya's agricultural policy :

This split has compromised the effective implementation of Paul Biya's agricultural policy, which aimed to promote an agricultural revolution to improve productivity, food security and the living conditions of farmers in Cameroon. Without close collaboration between research and agricultural extension, the ambitious objectives set by President Biya's agricultural policy have not been fully achieved. The rift between agricultural research and extension has had a negative impact on the implementation of Paul Biya's agricultural policy in Cameroon, compromising efforts to promote agricultural development and improve farmers' living conditions. It is therefore essential to strengthen collaboration between these two areas to ensure the success of agricultural initiatives and stimulate growth in the country's agricultural sector.

E - Incompatibility between academic theory and the pragmatic practice of agronomic training

The incompatibility between academic theory and the pragmatic practice of agronomic training has had a significant impact on the failure of agricultural development in Cameroon.

1. Academic theory versus pragmatic practice in agronomic training :

The agronomic training provided in universities and specialised schools often focuses on the theory and scientific concepts related to agriculture, such as cultivation techniques, soil management, seed selection, etc. However, this training does not place sufficient emphasis on the reality on the ground, the constraints encountered by local farmers, effective traditional practices, and the socio-economic and cultural aspects that influence the success of agriculture. However, these courses do not place sufficient emphasis on the reality on the ground, the constraints faced by local farmers, effective traditional practices, and the socio-economic and cultural aspects that influence the success of agriculture.

2. Impact on agricultural development in Cameroon :

The incompatibility between academic theory and the pragmatic practice of agronomic training has hindered agricultural development in Cameroon in several ways:

- Agronomists trained in academic methods may lack the practical skills to deal with the real challenges in the field, such as crop diseases, climate change or economic constraints.

- The agricultural innovations and technologies developed by researchers are not always adapted to local conditions and the needs of Cameroonian farmers, which limits their adoption and effectiveness.

- Ignorance of the socio-economic and cultural realities of rural communities can lead to inappropriate and ineffective interventions in agricultural development.

3. Consequences for the failure of agricultural development :

Because of this incompatibility between academic theory and the pragmatic practice of agronomic training, agricultural development in Cameroon has been hampered by several factors:

- Low agricultural productivity due to a lack of adoption of good farming practices and appropriate technologies.

-Inadequate food security and nutrition due to insufficient agricultural production and poor feeding practices.

- Persistent poverty in rural areas due to low agricultural income generation and limited access to markets.

4. Possible solutions to overcome this incompatibility :

To remedy this situation and promote agricultural development in Cameroon, it is essential to :

- Integrate more practical training and learning in the field into agronomy courses to enhance students' agricultural skills.

- Promote collaboration between universities, agricultural research centres, farmers' organisations and farmers to develop training programmes tailored to local needs.

- Encouraging applied research and participatory innovation to develop agricultural solutions adapted to the realities on the ground in Cameroon.

The incompatibility between academic theory and the pragmatic practice of agronomic training has contributed to the failure of agricultural development in Cameroon by limiting the effectiveness of agricultural interventions and hindering the progress of the agricultural sector. It is crucial to reform agronomic training programmes and promote a more holistic and participatory approach to overcome these obstacles and stimulate sustainable agricultural development in the country.

CHAPTER 4

PROSPECTS FOR PAUL BIYA'S AGRICULTURAL POLICY

A- Strategies to be put in place to strengthen the agricultural sector and achieve development objectives

Strengthening the agricultural sector and achieving development goals in this area requires the implementation of effective and coherent strategies. Here are some key strategies to consider in order to strengthen the agricultural sector and achieve development goals:

1. **Investing in agricultural infrastructure**: To improve agricultural productivity and stimulate growth in the sector, it is essential to invest in quality agricultural infrastructure, such as rural roads, irrigation networks, storage warehouses and agricultural markets. These infrastructures facilitate farmers' access to agricultural inputs, markets and services, thereby helping to strengthen the agricultural value chain.

2. **Promoting innovation and agricultural technology:** The adoption of innovative agricultural technologies can improve the efficiency and sustainability of agricultural production. It is therefore important to promote farmers' access to appropriate agricultural technologies, such as improved seeds, soil and water conservation techniques, and modern farming tools. Training farmers to use these technologies is also essential to maximise their impact.

3. **Building farmers' capacities**: To improve the performance of the agricultural sector, it is crucial to invest in building farmers' capacities

in terms of farm management, production techniques, marketing and risk management. Professional training and technical support can help farmers improve their skills and adopt more sustainable and profitable farming practices.

4. **Promoting access to agricultural finance:** Access to finance is a key element in the development of the agricultural sector. It is important to put in place financing mechanisms tailored to farmers' needs, such as agricultural credit at affordable interest rates, agricultural insurance and subsidy programmes for the purchase of agricultural inputs. These measures can help mitigate the financial risks associated with agriculture and stimulate investment in the sector.

5. **Promote collaboration and coordination between players in the agricultural sector:** To maximise the impact of agricultural interventions, it is important to promote collaboration and coordination between the various players in the sector, including farmers, professional organisations, government institutions, NGOs and the private sector. A participatory and inclusive approach can effectively mobilise the resources and skills needed to strengthen the agricultural sector and achieve development goals.

By implementing these key strategies, it is possible to strengthen the agricultural sector, improve food security, reduce rural poverty and contribute to sustainable economic development. It is therefore essential for governments, international organisations, private sector players and civil society to work together to put in place effective policies and programmes to promote agricultural and rural development.

B - Collaboration between national and international players to promote sustainable and inclusive agriculture

Collaboration between national and international players is essential to promote sustainable and inclusive agriculture in Cameroon. This collaboration makes it possible to draw on the expertise, resources and best practices available worldwide to strengthen the agricultural sector in the country. National stakeholders, such as the government, local farming organisations, cooperatives, agri-businesses and rural communities, play a crucial role in the development of sustainable agriculture in Cameroon. These actors are responsible for implementing agricultural policies, providing agricultural services to farmers, promoting sustainable agricultural practices and creating local markets for agricultural products. On the other hand, international players such as international organisations, NGOs, development agencies and donors provide financial, technical and strategic support to build the capacity of national players and promote sustainable agriculture in Cameroon. These international players can provide support for training farmers, providing access to quality seeds, setting up irrigation systems, promoting agroforestry, certifying agricultural products, setting up agricultural value chains and raising awareness of the importance of sustainable agriculture.By working together, national and international players can put in place innovative projects and programmes that help transform Cameroon's agricultural sector. For example, public-private partnerships can be established to invest in agricultural infrastructure, promote the use of cutting-edge agricultural technologies, encourage crop diversification, improve farmers' access to markets and support small-scale farmers.

Collaboration between national and international players is a fundamental pillar for promoting sustainable and inclusive agriculture in Cameroon. By pooling their efforts and resources, these players can help to strengthen food security, increase farmers' incomes, protect the environment and promote rural development in the country.

C - Expected impact of agricultural transformation on Cameroonian society and the national economy

The expected impact of agricultural transformation on Cameroonian society and the national economy is of great importance. Agriculture plays a crucial role in Cameroon's development, both socially and economically.

1. **Job creation**: Agricultural processing can help to create a large number of jobs, particularly in rural areas where agriculture is a main source of income. This could reduce unemployment and improve living conditions for local populations.

2. **Food security**: By modernising farming practices and improving yields, agricultural processing can help to boost food security in Cameroon. This would reduce dependence on food imports and guarantee adequate access to healthy food for the population.

3. **Higher incomes and poverty reduction**: More productive and profitable agriculture can lead to higher incomes for farmers and others involved in the agricultural sector. This can help to reduce poverty and socio-economic inequalities in the country.

4. Economic diversification: Agricultural processing can also contribute to the diversification of the Cameroonian economy by stimulating the development of related sectors such as agro-industry, agri-food processing and the export of processed agricultural products. This can strengthen the resilience of the national economy in the face of external shocks.

5. Preservation of the environment: It is essential that agricultural processing is carried out in a sustainable manner, taking into account the preservation of the environment. the environment and biodiversity. Environmentally-friendly farming practices can help combat climate change and preserve Cameroon's natural resources.

Agricultural transformation in Cameroon has great potential to have a positive impact on society and the national economy. However, it is essential to put in place effective policies and measures to support this process in an inclusive and sustainable way, involving local stakeholders and promoting resilient and environmentally-friendly agriculture.

D - Conclusion :

Agriculture is central to Paul Biya's vision of an emerging Cameroon by 2035. By investing in agricultural development, the Cameroonian government aims to stimulate economic growth, reduce poverty and improve the living conditions of rural populations. Continuing to support and strengthen the agricultural sector is essential to achieving this vision and making Cameroon a prosperous and sustainable emerging country. This book "Paul Biya, misunderstood architect of the agricultural revolution" highlights the key role played by President

Paul Biya in the modernisation and development of Cameroon's agricultural sector. Despite criticism and misunderstanding, Biya's commitment to improving farmers' living conditions and promoting sustainable agriculture remains undeniable. His legacy as the architect of the agricultural revolution will remain etched in Cameroon's history, and his work deserves to be recognised and celebrated for future generations.

REFERENCES

1. "Paul Biya: The Politics of Succession in Cameroon" by Willibroad Dze-Ngwa, African Books Collective, 2010.
2. "The Political Economy of Agricultural Policy Reform in Cameroon" by John Kuada, Nordic Africa Institute, 2002.
3. "Cameroon's Tycoon: Paul Biya and the Politics of Business in Cameroon" by Peter Ateh-Afac Fossungu, African Books Collective, 2019.
4. "Agricultural Development in Cameroon: Challenges and Prospects" by M. A. Nkeng, African Books Collective, 2011.
5. "Paul Biya and the Emergence of a Democratic State in Cameroon" by Victor Julius Ngoh, African Books Collective, 2004.
6. "Agricultural Policies in Sub-Saharan Africa: The Case of Cameroon" by Martin Ndende, African Books Collective, 2010.
7. "Biya's Cameroon: Politics and Stability in a Challenging Context" by Piet Konings and Francis B. Nyamnjoh, African Books Collective, 2019.
8. "Agricultural Transformation in Africa: The Case of Cameroon" by Martin Njeuma, African Books Collective, 2013.
9. "Paul Biya and the Struggle for Power in Cameroon" by Emmanuel M. Mbah, African Books Collective, 2004.
10. "Agricultural Development in Cameroon: Issues and Policy Options" by Emmanuel Nfor Nkimbeng, African Books Collective, 2014.

Printed by Books on Demand GmbH, Norderstedt / Germany